Space Voyager

Mercury

by Vanessa Black

Bullfrog Books

Ideas for Parents and Teachers

Bullfrog Books let children practice reading informational text at the earliest reading levels. Repetition, familiar words, and photo labels support early readers.

Before Reading

- Discuss the cover photo. What does it tell them?
- Look at the picture glossary together. Read and discuss the words.

Read the Book

- "Walk" through the book and look at the photos. Let the child ask questions. Point out the photo labels.
- Read the book to the child, or have him or her read independently.

After Reading

- Prompt the child to think more. Ask: Mercury is the closest planet to the sun. It is also very hot. What does this suggest to you about temperatures on other planets in our solar system?

Bullfrog Books are published by Jump!
5357 Penn Avenue South
Minneapolis, MN 55419
www.jumplibrary.com

Library of Congress Cataloging-in-Publication Data

Names: Black, Vanessa, 1973– author.
Title: Mercury / Vanessa Black.
Description: Minneapolis, MN: Jump!, Inc., [2018]
Series: Bullfrog Books. Space voyager
"Bullfrog Books are published by Jump!."
Audience: Ages 5–8. | Audience: K to grade 3.
Includes bibliographical references and index.
Identifiers: LCCN 2017021043 (print)
LCCN 2017021683 (ebook)
ISBN 9781624966866 (ebook)
ISBN 9781620318447 (hardcover: alk. paper)
ISBN 9781620318454 (pbk.)
Subjects: LCSH: Mercury (Planet)—Juvenile literature.
Classification: LCC QB611 (ebook)
LCC QB611 .B59 2017 (print) | DDC 523.41—dc23
LC record available at https://lccn.loc.gov/2017021043

Editor: Jenna Trnka
Book Designer: Molly Ballanger
Photo Researchers: Molly Ballanger & Jenna Trnka

Photo Credits: NASA images/Shutterstock, cover; s oleg/Shutterstock, 1 (boy); Nostalgia for Infinity/Shutterstock, 1 (drawing); StampCollection/Alamy, 3; Johns Hopkins University Applied Physics Laboratory/Carnegie Institution of Washington/NASA, 4, 19, 23tr; adventtr/iStock, 5, 23br; Mopic/Shutterstock, 6–7; Monica Schroeder/Science Source/Getty, 8–9, 23tl; NASA/Getty, 10–11; stevecoleimages/iStock, 12–13; Stocktrek Images/SuperStock, 14, 23ml; Science History Images/Alamy, 15; European Space Agency/Pierre Carril/Science Source, 16–17, 23mr; Andrey Popov/Shutterstock, 18; KK Tan/Shutterstock, 20; FlashMyPixel/iStock, 20–21; Aleksandr Pobedimskiy/Shutterstock, 23bl; 3Dsculptor/Shutterstock, 24.

Printed in the United States of America at Corporate Graphics in North Mankato, Minnesota.

Table of Contents

Small and Hot 4
A Look at Mercury 22
Picture Glossary 23
Index 24
To Learn More 24

Small and Hot

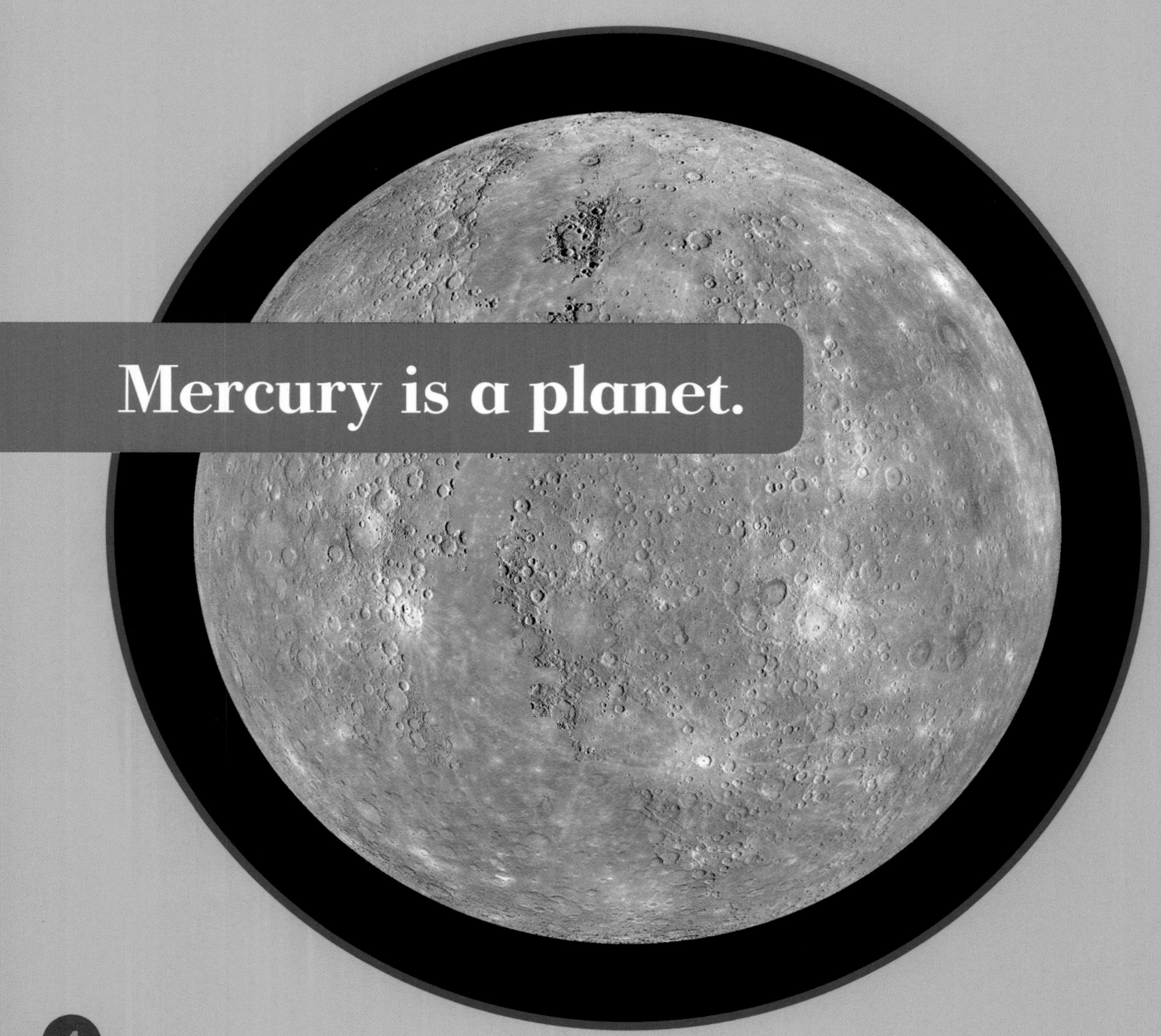

Mercury is a planet.

It is in our solar system!

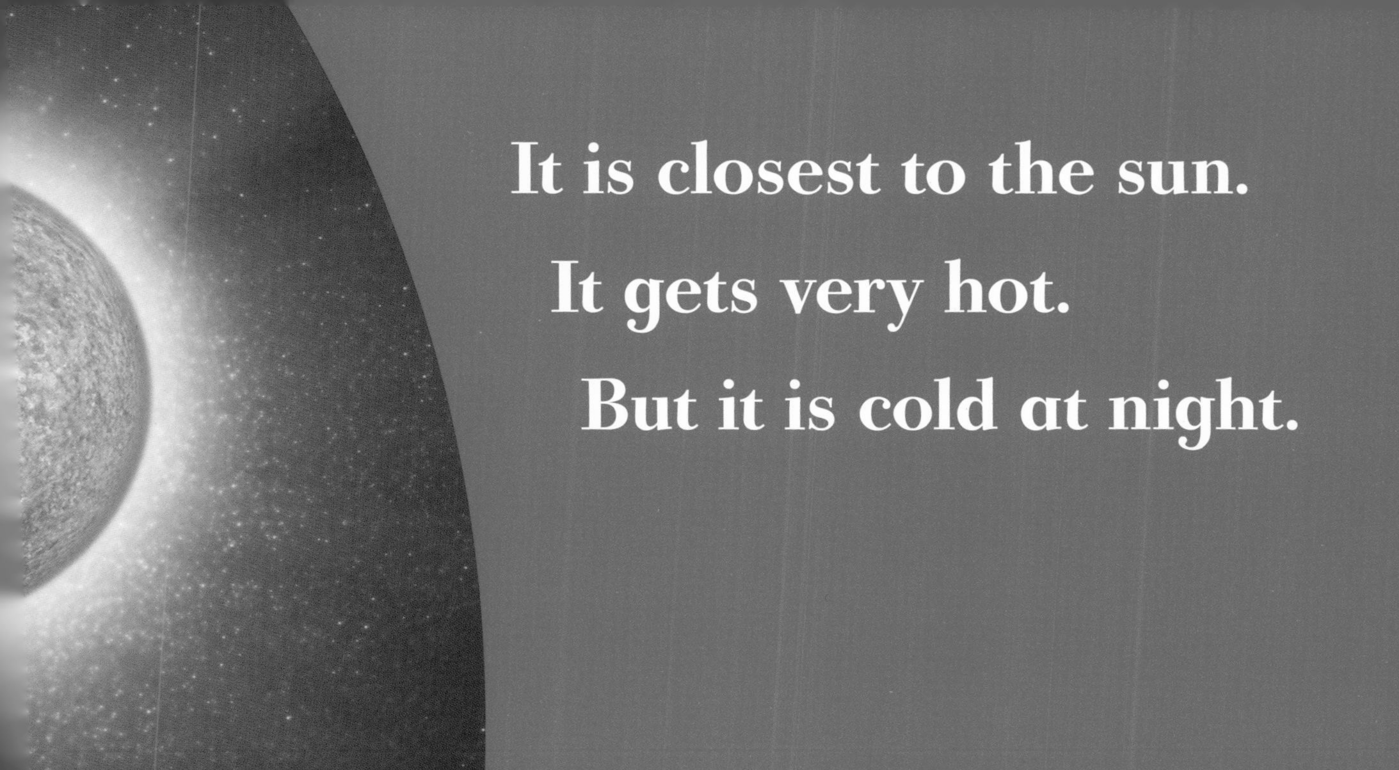

It is closest to the sun.

It gets very hot.

But it is cold at night.

Mercury is small.

But it has a big core.

It is iron.

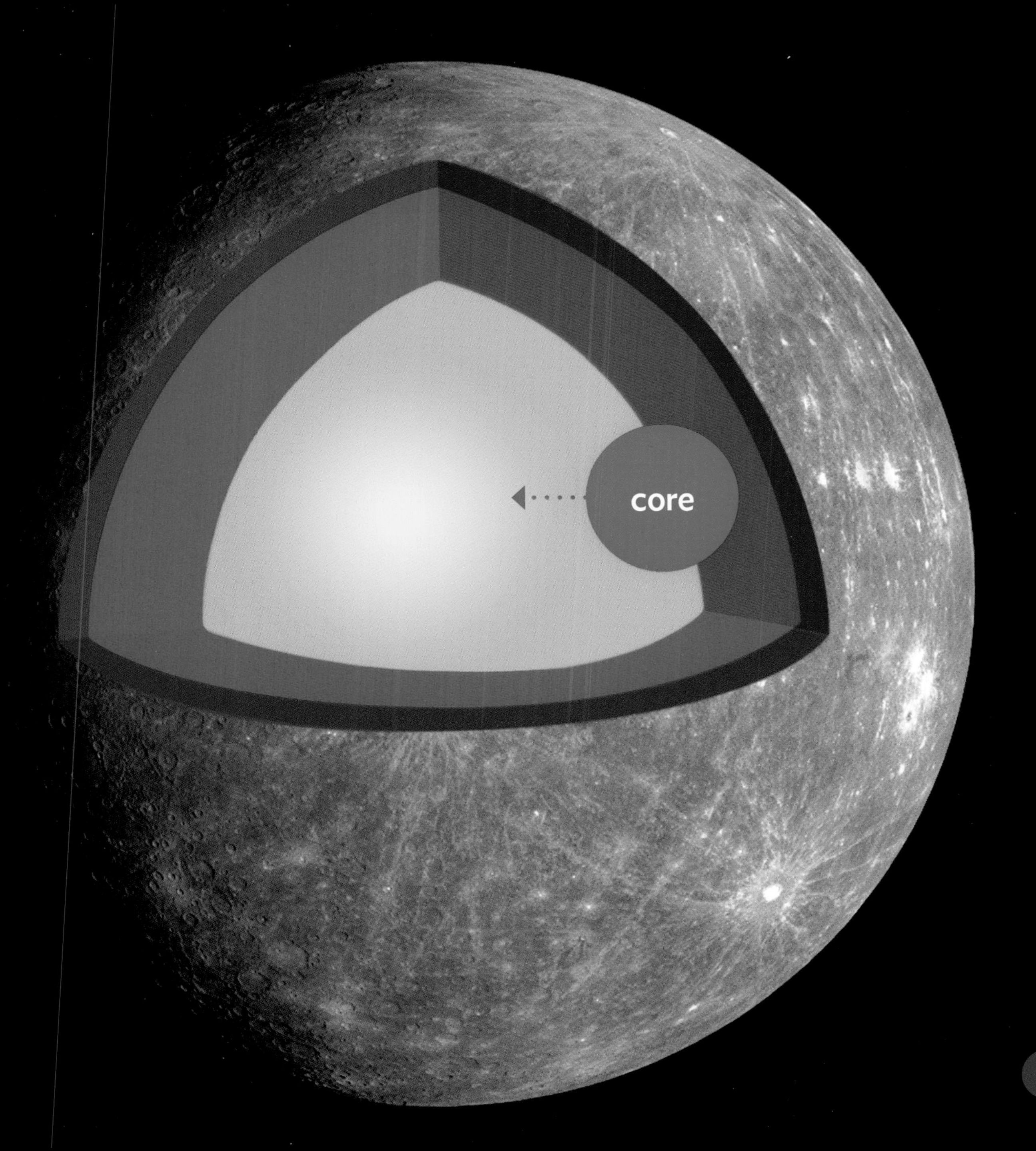
core

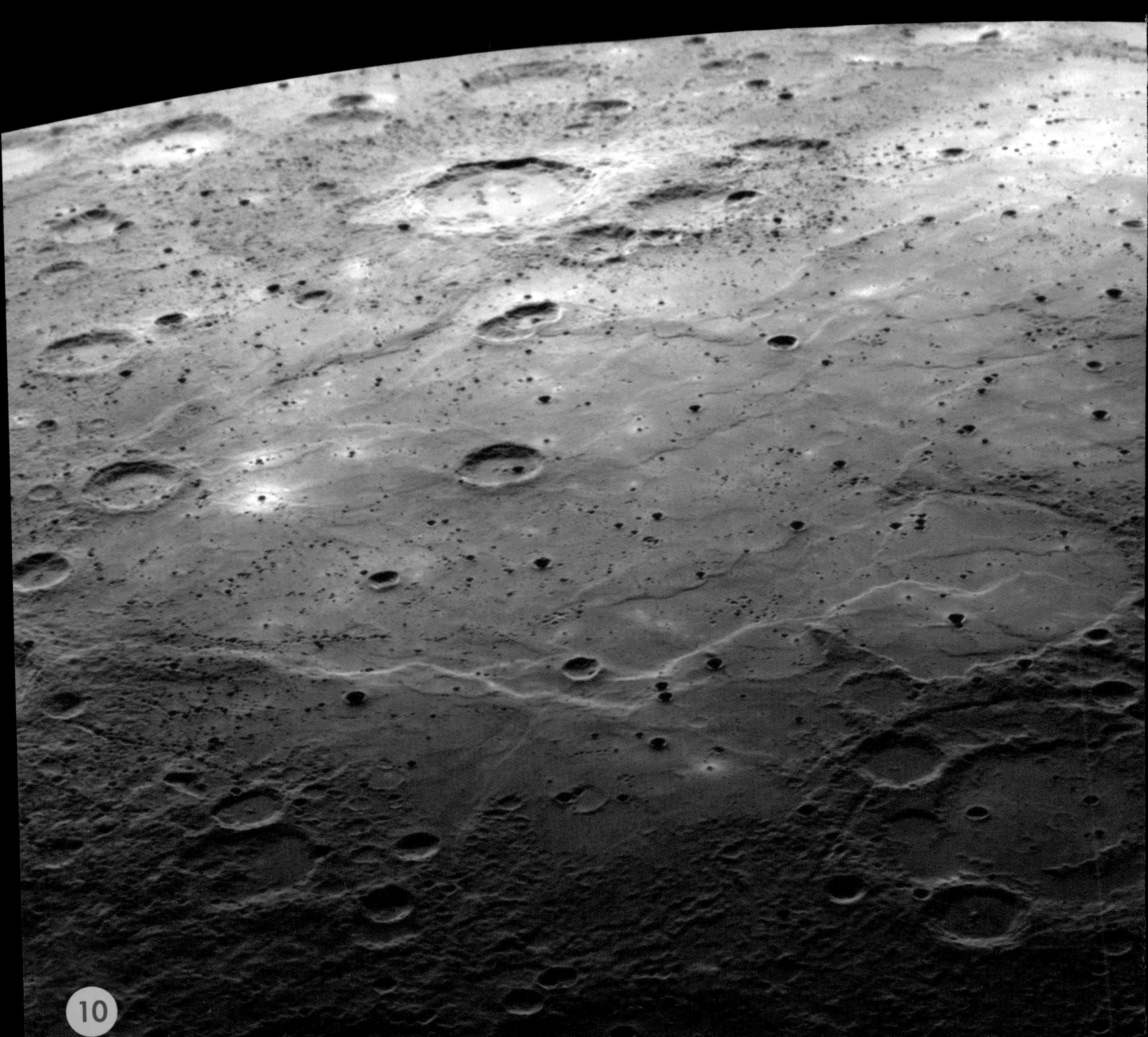
10

It is a gray color.
It has rocks.

How do we know?
We use a telescope.
We look at it.

telescope

Look! What do we see?

Big craters.

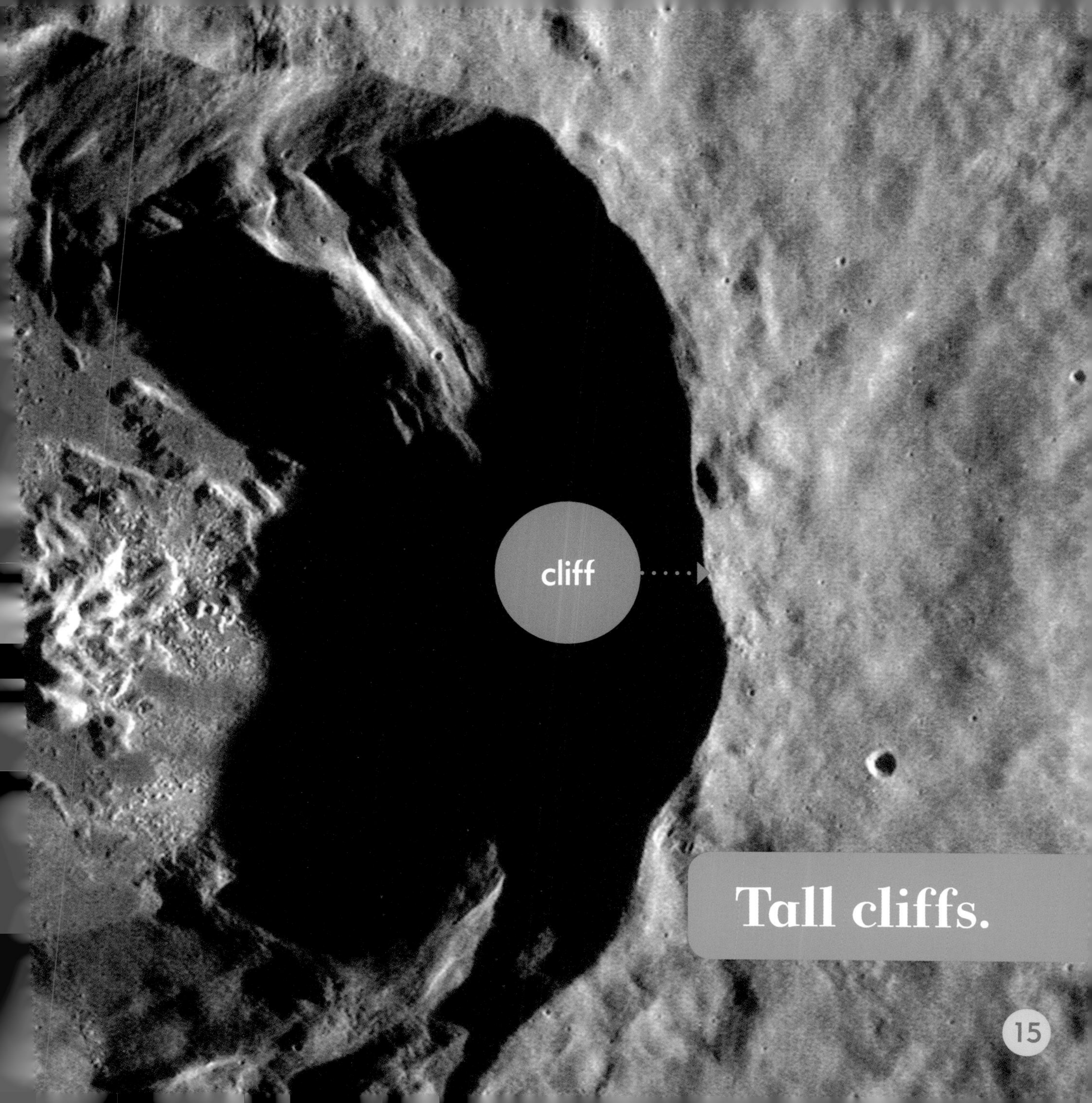

Tall cliffs.

We send out probes.
They take photos.

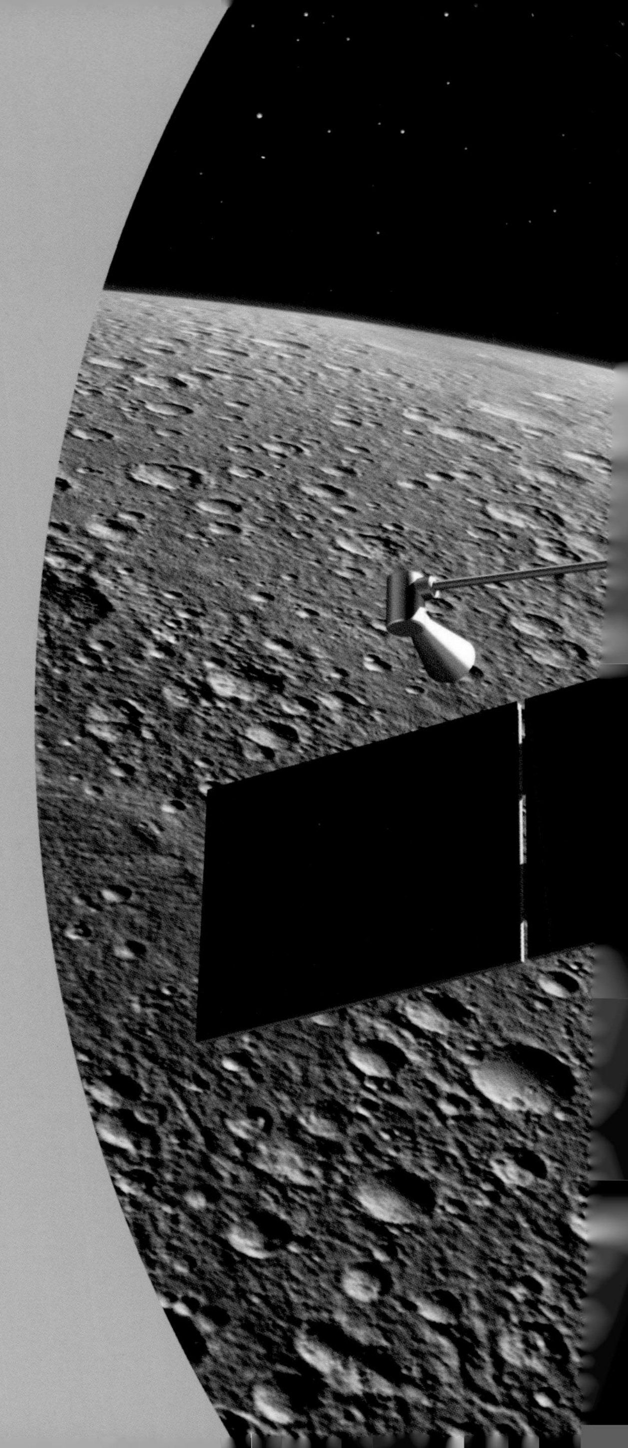

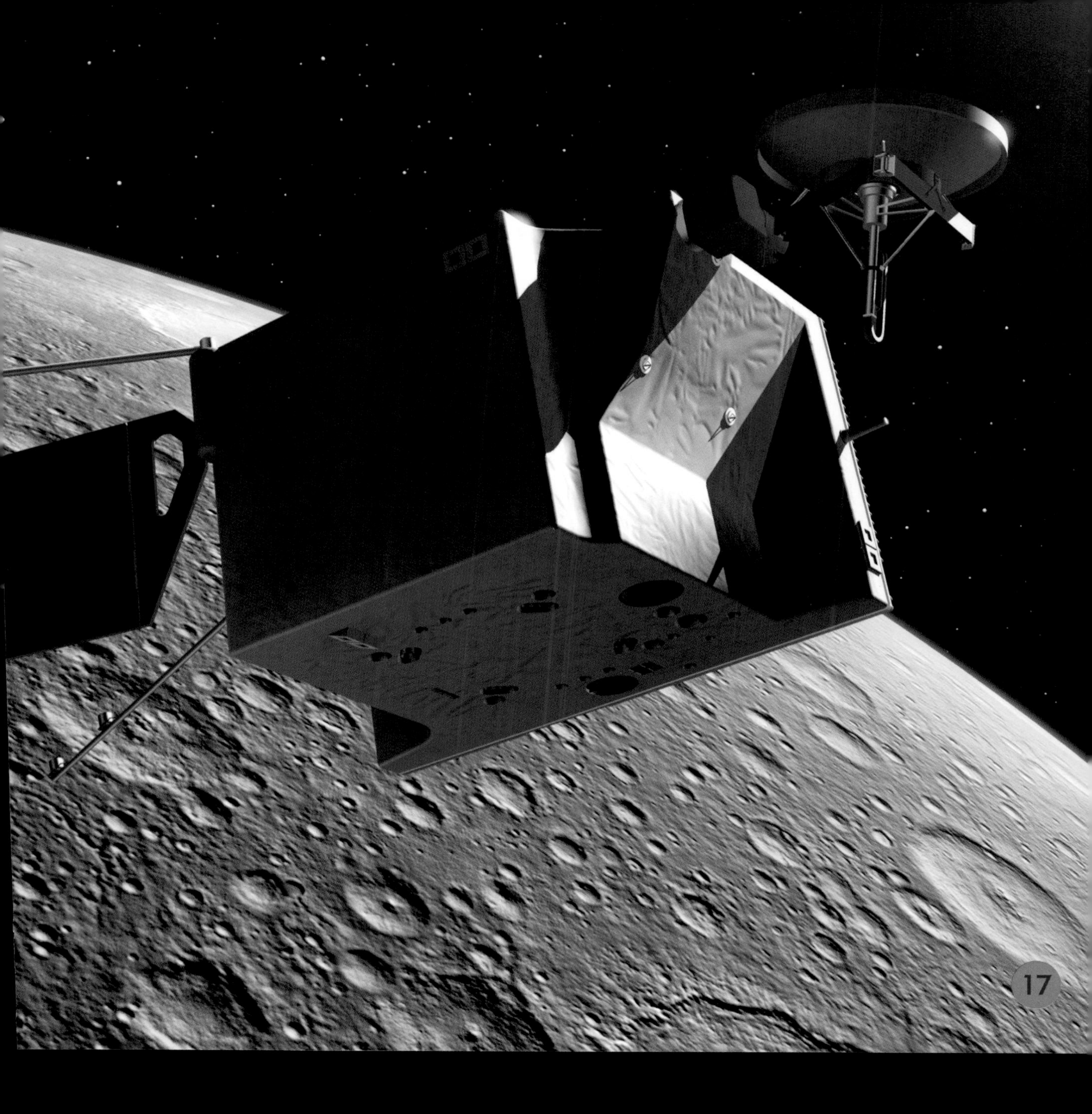

What do they show?

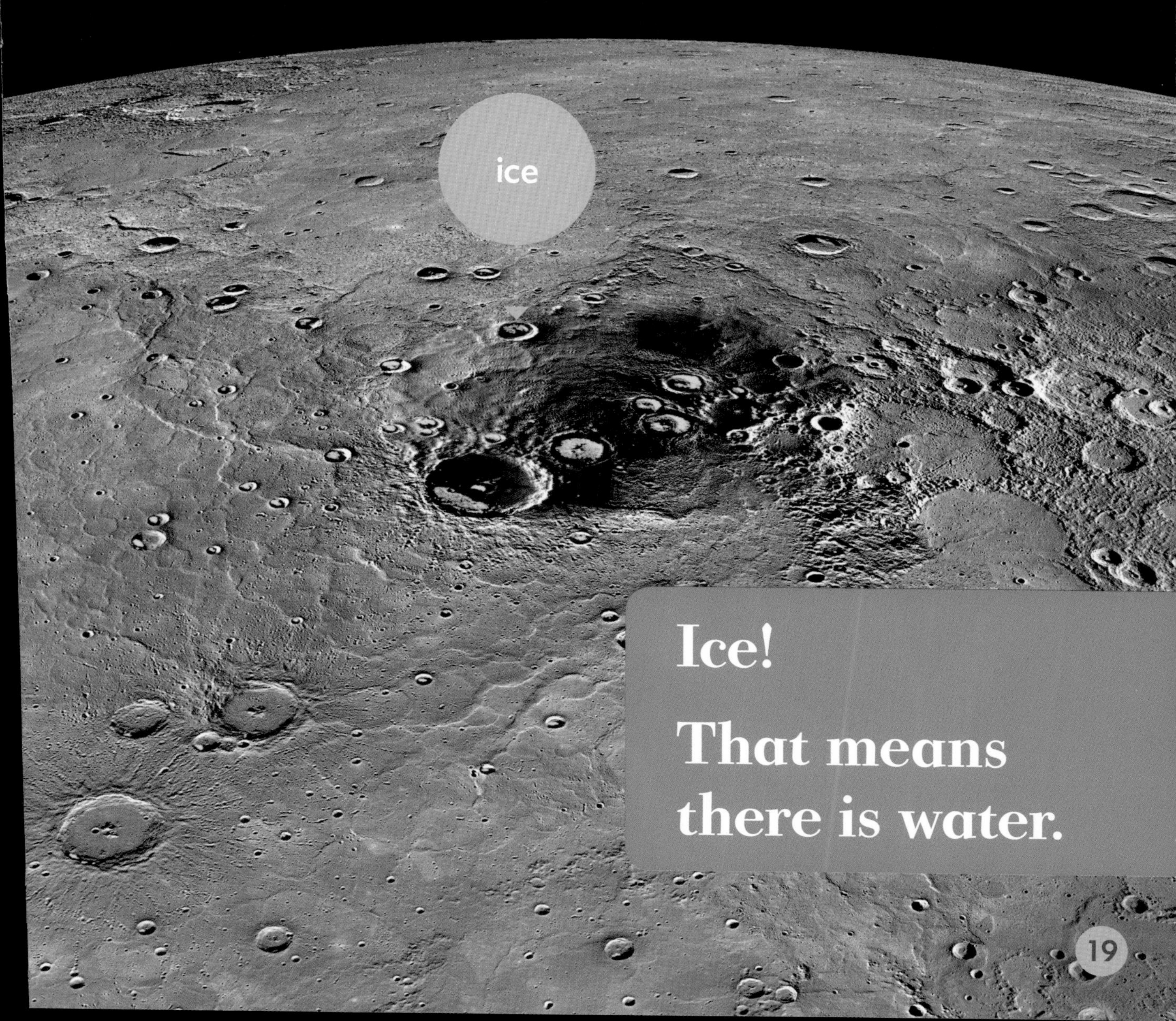

Ice!

That means there is water.

Wow!
Why do you like Mercury?

A Look at Mercury

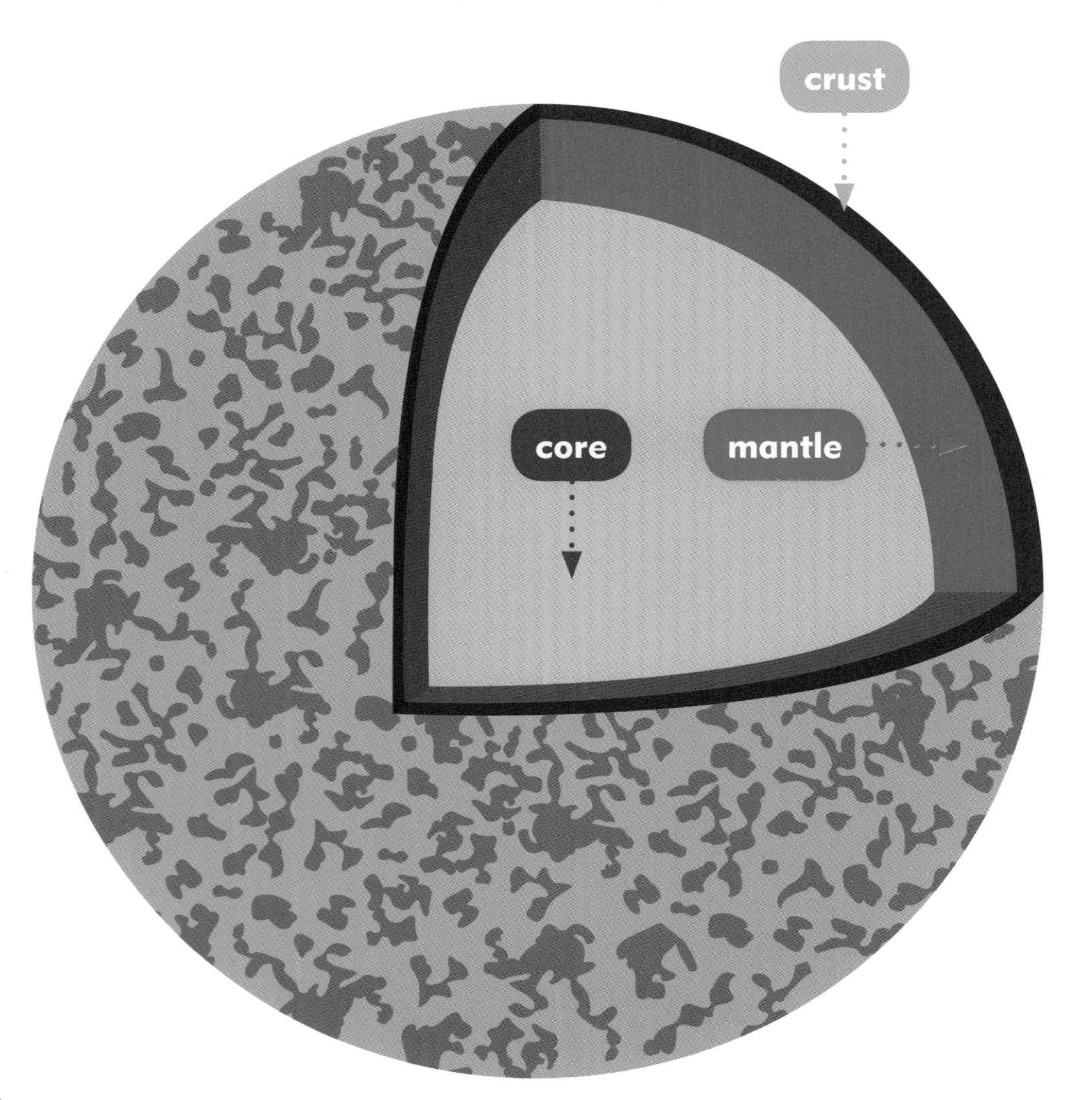

Picture Glossary

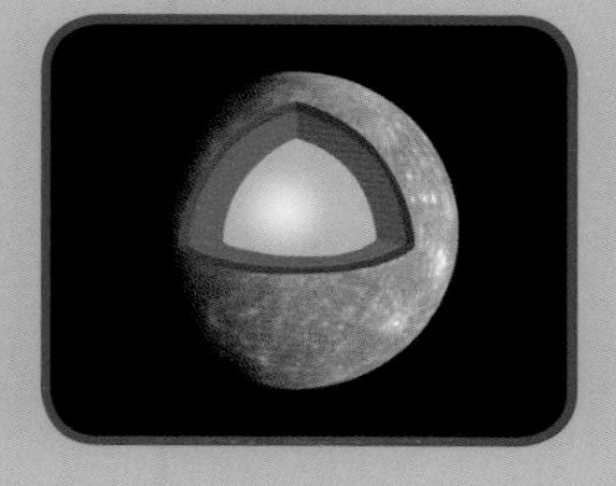

core
The inner part.

craters
Large round holes.

iron
A heavy metal found in nature.

planet
A large body that orbits the sun.

probes
Spacecraft used to explore space.

solar system
The sun and other planets that revolve around it.

Index

cliffs 15
core 8
craters 14
gray 11
ice 19
iron 8
planet 4
probes 16
rocks 11
solar system 5
sun 7
telescope 12

To Learn More

Learning more is as easy as 1, 2, 3.

1) Go to www.factsurfer.com

2) Enter "Mercury" into the search box.

3) Click the "Surf" button to see a list of websites.

With factsurfer.com, finding more information is just a click away.